Titles in This Series

My Alphabet Book
My Counting Book
My Book of Colors and Shapes
My Book of Opposites

LADYBIRD BOOKS, INC.
Auburn, Maine 04210 U.S.A.
© LADYBIRD BOOKS LTD MCMLXXXVIII
Loughborough, Leicestershire, England

Printed in England

My Counting Book

by RONNE PELTZMAN RANDALL
illustrated by STEVE SMALLMAN

Ladybird Books

1 One rabbit home alone.

Two rabbits on the phone.

Three rabbits at the door.

4 **Four** rabbits at the store.

Special Offer 29¢

5 **Five** rabbits mix and bake.

6

Six rabbits have a cake!

7

Seven rabbits with plates
and spoons.

8

Eight rabbits with bright balloons.

POP!

Nine rabbits' work is done.

10

Now **ten** rabbits have some fun!

How many rabbits came
to the party?

Can you find each of these rabbits
in the pictures in this book?